The Standard®

ENGINE LOG

ASA-SE-2

THE STANDARD® ENGINE LOG
SE-2

ASA-SE-2
ISBN 978-1-56027-326-4

Published by
Aviation Supplies & Academics, Inc.
7005 132nd Place SE
Newcastle, Washington 98059 USA
Website: asa2fly.com
Email: asa@asa2fly.com

Printed in Malaysia

[18] 22

TRANSPORTATION USD $14.95

ISBN 978-1-56027-326-4

51495 >

9 781560 273264

Engine Record General Information

Manufacturer ______________________ **Model** ______________________

Serial ______________________ **Type Certificate** ______________________

This engine is currently installed in aircraft: ______________________

____

Minimum Octane Fuel ____________ **Oil Grade** Summer ____________ Winter ____________

Magneto Time ____________ **Point Setting** ____________ **Firing Order** ____________

Spark Plug Gap ____________

Manufacturer's Recommended Overhaul at ____________ **hours**

Registered Owner Record

Name ______________________ Address ______________________

City ______________ State ________ From ______________ To ______________

Name ______________________ Address ______________________

City ______________ State ________ From ______________ To ______________

Name ______________________ Address ______________________

City ______________ State ________ From ______________ To ______________

Name ______________________ Address ______________________

City ______________ State ________ From ______________ To ______________

Name ______________________ Address ______________________

City ______________ State ________ From ______________ To ______________

Name ______________________ Address ______________________

City ______________ State ________ From ______________ To ______________

Name ______________________ Address ______________________

City ______________ State ________ From ______________ To ______________

Name ______________________ Address ______________________

City ______________ State ________ From ______________ To ______________

Name ______________________ Address ______________________

City ______________ State ________ From ______________ To ______________

Notes

YEAR 20 ____ DATE	RECORDING TACH TIME	TODAY'S FLIGHT	TOTAL TIME IN SERVICE	**Description of Inspections, Tests, Repairs, and Alterations** Entries must be endorsed with Name, Rating and Certificate Number of Technician or Repair Facility. (See back pages for other specific entries.)

EAR O ____ ATE	RECORDING TACH TIME	TODAY'S FLIGHT	TOTAL TIME IN SERVICE	**Description of Inspections, Tests, Repairs, and Alterations** Entries must be endorsed with Name, Rating and Certificate Number of Technician or Repair Facility. (See back pages for other specific entries.)

YEAR 20 ____ DATE	RECORDING TACH TIME	TODAY'S FLIGHT	TOTAL TIME IN SERVICE	**Description of Inspections, Tests, Repairs, and Alterations** Entries must be endorsed with Name, Rating and Certificate Number of Technician or Repair Facility. (See back pages for other specific entries.)

YEAR 20 ____ DATE	RECORDING TACH TIME	TODAY'S FLIGHT	TOTAL TIME IN SERVICE	**Description of Inspections, Tests, Repairs, and Alterations** Entries must be endorsed with Name, Rating and Certificate Number of Technician or Repair Facility. (See back pages for other specific entries.)

YEAR 20 ______ DATE	RECORDING TACH TIME	TODAY'S FLIGHT	TOTAL TIME IN SERVICE	**Description of Inspections, Tests, Repairs, and Alterations** Entries must be endorsed with Name, Rating and Certificate Number of Technician or Repair Facility. (See back pages for other specific entries.)

YEAR 20 ____ DATE	RECORDING TACH TIME	TODAY'S FLIGHT	TOTAL TIME IN SERVICE	**Description of Inspections, Tests, Repairs, and Alterations** Entries must be endorsed with Name, Rating and Certificate Number of Technician or Repair Facility. (See back pages for other specific entries.)

YEAR 20 ____ DATE	RECORDING TACH TIME	TODAY'S FLIGHT	TOTAL TIME IN SERVICE	**Description of Inspections, Tests, Repairs, and Alterations** Entries must be endorsed with Name, Rating and Certificate Number of Technician or Repair Facility. (See back pages for other specific entries.)

'EAR 0 ____ ATE	RECORDING TACH TIME	TODAY'S FLIGHT	TOTAL TIME IN SERVICE	**Description of Inspections, Tests, Repairs, and Alterations** Entries must be endorsed with Name, Rating and Certificate Number of Technician or Repair Facility. (See back pages for other specific entries.)

YEAR 20 _____ DATE	RECORDING TACH TIME	TODAY'S FLIGHT	TOTAL TIME IN SERVICE	**Description of Inspections, Tests, Repairs, and Alterations** Entries must be endorsed with Name, Rating and Certificate Number of Technician or Repair Facility. (See back pages for other specific entries.)

'EAR 0 ____ ATE	RECORDING TACH TIME	TODAY'S FLIGHT	TOTAL TIME IN SERVICE	**Description of Inspections, Tests, Repairs, and Alterations** Entries must be endorsed with Name, Rating and Certificate Number of Technician or Repair Facility. (See back pages for other specific entries.)

YEAR 20____ DATE	RECORDING TACH TIME	TODAY'S FLIGHT	TOTAL TIME IN SERVICE	**Description of Inspections, Tests, Repairs, and Alterations** Entries must be endorsed with Name, Rating and Certificate Number of Technician or Repair Facility. (See back pages for other specific entries.)

EAR O ______ ATE	RECORDING TACH TIME	TODAY'S FLIGHT	TOTAL TIME IN SERVICE	**Description of Inspections, Tests, Repairs, and Alterations** Entries must be endorsed with Name, Rating and Certificate Number of Technician or Repair Facility. (See back pages for other specific entries.)

YEAR 20 ____ DATE	RECORDING TACH TIME	TODAY'S FLIGHT	TOTAL TIME IN SERVICE	**Description of Inspections, Tests, Repairs, and Alterations** Entries must be endorsed with Name, Rating and Certificate Number of Technician or Repair Facility. (See back pages for other specific entries.)

EAR O ____ ATE	RECORDING TACH TIME	TODAY'S FLIGHT	TOTAL TIME IN SERVICE	**Description of Inspections, Tests, Repairs, and Alterations** Entries must be endorsed with Name, Rating and Certificate Number of Technician or Repair Facility. (See back pages for other specific entries.)

YEAR 20 ____ DATE	RECORDING TACH TIME	TODAY'S FLIGHT	TOTAL TIME IN SERVICE	**Description of Inspections, Tests, Repairs, and Alterations** Entries must be endorsed with Name, Rating and Certificate Number of Technician or Repair Facility. (See back pages for other specific entries.)

EAR ATE	RECORDING TACH TIME	TODAY'S FLIGHT	TOTAL TIME IN SERVICE	**Description of Inspections, Tests, Repairs, and Alterations** Entries must be endorsed with Name, Rating and Certificate Number of Technician or Repair Facility. (See back pages for other specific entries.)

YEAR 20 ____ DATE	RECORDING TACH TIME	TODAY'S FLIGHT	TOTAL TIME IN SERVICE	**Description of Inspections, Tests, Repairs, and Alterations** Entries must be endorsed with Name, Rating and Certificate Number of Technician or Repair Facility. (See back pages for other specific entries.)

EAR ____ ATE	RECORDING TACH TIME	TODAY'S FLIGHT	TOTAL TIME IN SERVICE	**Description of Inspections, Tests, Repairs, and Alterations** Entries must be endorsed with Name, Rating and Certificate Number of Technician or Repair Facility. (See back pages for other specific entries.)

YEAR 20 ____ DATE	RECORDING TACH TIME	TODAY'S FLIGHT	TOTAL TIME IN SERVICE	**Description of Inspections, Tests, Repairs, and Alterations** Entries must be endorsed with Name, Rating and Certificate Number of Technician or Repair Facility. (See back pages for other specific entries.)

YEAR 20 _____ DATE	RECORDING TACH TIME	TODAY'S FLIGHT	TOTAL TIME IN SERVICE	**Description of Inspections, Tests, Repairs, and Alterations** Entries must be endorsed with Name, Rating and Certificate Number of Technician or Repair Facility. (See back pages for other specific entries.)

YEAR 20 ____ DATE	RECORDING TACH TIME	TODAY'S FLIGHT	TOTAL TIME IN SERVICE	**Description of Inspections, Tests, Repairs, and Alterations** Entries must be endorsed with Name, Rating and Certificate Number of Technician or Repair Facility. (See back pages for other specific entries.)

YEAR 20 ____ DATE	RECORDING TACH TIME	TODAY'S FLIGHT	TOTAL TIME IN SERVICE	**Description of Inspections, Tests, Repairs, and Alterations** Entries must be endorsed with Name, Rating and Certificate Number of Technician or Repair Facility. (See back pages for other specific entries.)

YEAR 20 _____ DATE	RECORDING TACH TIME	TODAY'S FLIGHT	TOTAL TIME IN SERVICE	**Description of Inspections, Tests, Repairs, and Alterations** Entries must be endorsed with Name, Rating and Certificate Number of Technician or Repair Facility. (See back pages for other specific entries.)

YEAR 20____ DATE	RECORDING TACH TIME	TODAY'S FLIGHT	TOTAL TIME IN SERVICE	**Description of Inspections, Tests, Repairs, and Alterations** Entries must be endorsed with Name, Rating and Certificate Number of Technician or Repair Facility. (See back pages for other specific entries.)

YEAR 20 ____ DATE	RECORDING TACH TIME	TODAY'S FLIGHT	TOTAL TIME IN SERVICE	**Description of Inspections, Tests, Repairs, and Alterations** Entries must be endorsed with Name, Rating and Certificate Number of Technician or Repair Facility. (See back pages for other specific entries.)

EAR 0 ____ ATE	RECORDING TACH TIME	TODAY'S FLIGHT	TOTAL TIME IN SERVICE	**Description of Inspections, Tests, Repairs, and Alterations** Entries must be endorsed with Name, Rating and Certificate Number of Technician or Repair Facility. (See back pages for other specific entries.)

YEAR 20 ____ DATE	RECORDING TACH TIME	TODAY'S FLIGHT	TOTAL TIME IN SERVICE	**Description of Inspections, Tests, Repairs, and Alterations** Entries must be endorsed with Name, Rating and Certificate Number of Technician or Repair Facility. (See back pages for other specific entries.)

YEAR 20 _____ DATE	RECORDING TACH TIME	TODAY'S FLIGHT	TOTAL TIME IN SERVICE	**Description of Inspections, Tests, Repairs, and Alterations** Entries must be endorsed with Name, Rating and Certificate Number of Technician or Repair Facility. (See back pages for other specific entries.)

YEAR 20____ DATE	RECORDING TACH TIME	TODAY'S FLIGHT	TOTAL TIME IN SERVICE	**Description of Inspections, Tests, Repairs, and Alterations** Entries must be endorsed with Name, Rating and Certificate Number of Technician or Repair Facility. (See back pages for other specific entries.)

YEAR 20 ____ DATE	RECORDING TACH TIME	TODAY'S FLIGHT	TOTAL TIME IN SERVICE	**Description of Inspections, Tests, Repairs, and Alterations** Entries must be endorsed with Name, Rating and Certificate Number of Technician or Repair Facility. (See back pages for other specific entries.)

YEAR 20 ____ DATE	RECORDING TACH TIME	TODAY'S FLIGHT	TOTAL TIME IN SERVICE	**Description of Inspections, Tests, Repairs, and Alterations** Entries must be endorsed with Name, Rating and Certificate Number of Technician or Repair Facility. (See back pages for other specific entries.)

YEAR 20 ____ DATE	RECORDING TACH TIME	TODAY'S FLIGHT	TOTAL TIME IN SERVICE	**Description of Inspections, Tests, Repairs, and Alterations** Entries must be endorsed with Name, Rating and Certificate Number of Technician or Repair Facility. (See back pages for other specific entries.)

YEAR 20 ____ DATE	RECORDING TACH TIME	TODAY'S FLIGHT	TOTAL TIME IN SERVICE	**Description of Inspections, Tests, Repairs, and Alterations** Entries must be endorsed with Name, Rating and Certificate Number of Technician or Repair Facility. (See back pages for other specific entries.)

YEAR 20____ DATE	RECORDING TACH TIME	TODAY'S FLIGHT	TOTAL TIME IN SERVICE	**Description of Inspections, Tests, Repairs, and Alterations** Entries must be endorsed with Name, Rating and Certificate Number of Technician or Repair Facility. (See back pages for other specific entries.)

YEAR 20 ____ DATE	RECORDING TACH TIME	TODAY'S FLIGHT	TOTAL TIME IN SERVICE	**Description of Inspections, Tests, Repairs, and Alterations** Entries must be endorsed with Name, Rating and Certificate Number of Technician or Repair Facility. (See back pages for other specific entries.)

EAR 0 ____ ATE	RECORDING TACH TIME	TODAY'S FLIGHT	TOTAL TIME IN SERVICE	**Description of Inspections, Tests, Repairs, and Alterations** Entries must be endorsed with Name, Rating and Certificate Number of Technician or Repair Facility. (See back pages for other specific entries.)

YEAR 20 _____ DATE	RECORDING TACH TIME	TODAY'S FLIGHT	TOTAL TIME IN SERVICE	**Description of Inspections, Tests, Repairs, and Alterations** Entries must be endorsed with Name, Rating and Certificate Number of Technician or Repair Facility. (See back pages for other specific entries.)

YEAR 20____ DATE	RECORDING TACH TIME	TODAY'S FLIGHT	TOTAL TIME IN SERVICE	**Description of Inspections, Tests, Repairs, and Alterations** Entries must be endorsed with Name, Rating and Certificate Number of Technician or Repair Facility. (See back pages for other specific entries.)

YEAR 20 ____ DATE	RECORDING TACH TIME	TODAY'S FLIGHT	TOTAL TIME IN SERVICE	**Description of Inspections, Tests, Repairs, and Alterations** Entries must be endorsed with Name, Rating and Certificate Number of Technician or Repair Facility. (See back pages for other specific entries.)

EAR O ____ ATE	RECORDING TACH TIME	TODAY'S FLIGHT	TOTAL TIME IN SERVICE	**Description of Inspections, Tests, Repairs, and Alterations** Entries must be endorsed with Name, Rating and Certificate Number of Technician or Repair Facility. (See back pages for other specific entries.)

YEAR 20 _____ DATE	RECORDING TACH TIME	TODAY'S FLIGHT	TOTAL TIME IN SERVICE	**Description of Inspections, Tests, Repairs, and Alterations** Entries must be endorsed with Name, Rating and Certificate Number of Technician or Repair Facility. (See back pages for other specific entries.)

EAR 0 ______ ATE	RECORDING TACH TIME	TODAY'S FLIGHT	TOTAL TIME IN SERVICE	**Description of Inspections, Tests, Repairs, and Alterations** Entries must be endorsed with Name, Rating and Certificate Number of Technician or Repair Facility. (See back pages for other specific entries.)

YEAR 20 ____ DATE	RECORDING TACH TIME	TODAY'S FLIGHT	TOTAL TIME IN SERVICE	**Description of Inspections, Tests, Repairs, and Alterations** Entries must be endorsed with Name, Rating and Certificate Number of Technician or Repair Facility. (See back pages for other specific entries.)

YEAR 20 ____ DATE	RECORDING TACH TIME	TODAY'S FLIGHT	TOTAL TIME IN SERVICE	**Description of Inspections, Tests, Repairs, and Alterations** Entries must be endorsed with Name, Rating and Certificate Number of Technician or Repair Facility. (See back pages for other specific entries.)

YEAR 20 _____ DATE	RECORDING TACH TIME	TODAY'S FLIGHT	TOTAL TIME IN SERVICE	**Description of Inspections, Tests, Repairs, and Alterations** Entries must be endorsed with Name, Rating and Certificate Number of Technician or Repair Facility. (See back pages for other specific entries.)

YEAR 20 ____ DATE	RECORDING TACH TIME	TODAY'S FLIGHT	TOTAL TIME IN SERVICE	**Description of Inspections, Tests, Repairs, and Alterations** Entries must be endorsed with Name, Rating and Certificate Number of Technician or Repair Facility. (See back pages for other specific entries.)

YEAR 20 ____ DATE	RECORDING TACH TIME	TODAY'S FLIGHT	TOTAL TIME IN SERVICE	**Description of Inspections, Tests, Repairs, and Alterations** Entries must be endorsed with Name, Rating and Certificate Number of Technician or Repair Facility. (See back pages for other specific entries.)

YEAR 20 ____ DATE	RECORDING TACH TIME	TODAY'S FLIGHT	TOTAL TIME IN SERVICE	**Description of Inspections, Tests, Repairs, and Alterations** Entries must be endorsed with Name, Rating and Certificate Number of Technician or Repair Facility. (See back pages for other specific entries.)

YEAR 20 ____ DATE	RECORDING TACH TIME	TODAY'S FLIGHT	TOTAL TIME IN SERVICE	**Description of Inspections, Tests, Repairs, and Alterations** Entries must be endorsed with Name, Rating and Certificate Number of Technician or Repair Facility. (See back pages for other specific entries.)

YEAR 20 ____ DATE	RECORDING TACH TIME	TODAY'S FLIGHT	TOTAL TIME IN SERVICE	**Description of Inspections, Tests, Repairs, and Alterations** Entries must be endorsed with Name, Rating and Certificate Number of Technician or Repair Facility. (See back pages for other specific entries.)

YEAR 20 ____ DATE	RECORDING TACH TIME	TODAY'S FLIGHT	TOTAL TIME IN SERVICE	**Description of Inspections, Tests, Repairs, and Alterations** Entries must be endorsed with Name, Rating and Certificate Number of Technician or Repair Facility. (See back pages for other specific entries.)

YEAR 20 ____ DATE	RECORDING TACH TIME	TODAY'S FLIGHT	TOTAL TIME IN SERVICE	**Description of Inspections, Tests, Repairs, and Alterations** Entries must be endorsed with Name, Rating and Certificate Number of Technician or Repair Facility. (See back pages for other specific entries.)

YEAR 20 ____ DATE	RECORDING TACH TIME	TODAY'S FLIGHT	TOTAL TIME IN SERVICE	**Description of Inspections, Tests, Repairs, and Alterations** Entries must be endorsed with Name, Rating and Certificate Number of Technician or Repair Facility. (See back pages for other specific entries.)

YEAR 20____ DATE	RECORDING TACH TIME	TODAY'S FLIGHT	TOTAL TIME IN SERVICE	**Description of Inspections, Tests, Repairs, and Alterations** Entries must be endorsed with Name, Rating and Certificate Number of Technician or Repair Facility. (See back pages for other specific entries.)

YEAR 20 ____ DATE	RECORDING TACH TIME	TODAY'S FLIGHT	TOTAL TIME IN SERVICE	**Description of Inspections, Tests, Repairs, and Alterations** Entries must be endorsed with Name, Rating and Certificate Number of Technician or Repair Facility. (See back pages for other specific entries.)

EAR 0 _____ ATE	RECORDING TACH TIME	TODAY'S FLIGHT	TOTAL TIME IN SERVICE	**Description of Inspections, Tests, Repairs, and Alterations** Entries must be endorsed with Name, Rating and Certificate Number of Technician or Repair Facility. (See back pages for other specific entries.)

YEAR 20 ____ DATE	RECORDING TACH TIME	TODAY'S FLIGHT	TOTAL TIME IN SERVICE	**Description of Inspections, Tests, Repairs, and Alterations** Entries must be endorsed with Name, Rating and Certificate Number of Technician or Repair Facility. (See back pages for other specific entries.)

EAR O ______ ATE	RECORDING TACH TIME	TODAY'S FLIGHT	TOTAL TIME IN SERVICE	**Description of Inspections, Tests, Repairs, and Alterations** Entries must be endorsed with Name, Rating and Certificate Number of Technician or Repair Facility. (See back pages for other specific entries.)

YEAR 20 _____ DATE	RECORDING TACH TIME	TODAY'S FLIGHT	TOTAL TIME IN SERVICE	**Description of Inspections, Tests, Repairs, and Alterations** Entries must be endorsed with Name, Rating and Certificate Number of Technician or Repair Facility. (See back pages for other specific entries.)

YEAR 20 ____ DATE	RECORDING TACH TIME	TODAY'S FLIGHT	TOTAL TIME IN SERVICE	**Description of Inspections, Tests, Repairs, and Alterations** Entries must be endorsed with Name, Rating and Certificate Number of Technician or Repair Facility. (See back pages for other specific entries.)

YEAR 20 _____ DATE	RECORDING TACH TIME	TODAY'S FLIGHT	TOTAL TIME IN SERVICE	**Description of Inspections, Tests, Repairs, and Alterations** Entries must be endorsed with Name, Rating and Certificate Number of Technician or Repair Facility. (See back pages for other specific entries.)

YEAR 20 _____ DATE	RECORDING TACH TIME	TODAY'S FLIGHT	TOTAL TIME IN SERVICE	**Description of Inspections, Tests, Repairs, and Alterations** Entries must be endorsed with Name, Rating and Certificate Number of Technician or Repair Facility. (See back pages for other specific entries.)

YEAR 20 ____ DATE	RECORDING TACH TIME	TODAY'S FLIGHT	TOTAL TIME IN SERVICE	**Description of Inspections, Tests, Repairs, and Alterations** Entries must be endorsed with Name, Rating and Certificate Number of Technician or Repair Facility. (See back pages for other specific entries.)

YEAR 20 ____ DATE	RECORDING TACH TIME	TODAY'S FLIGHT	TOTAL TIME IN SERVICE	**Description of Inspections, Tests, Repairs, and Alterations** Entries must be endorsed with Name, Rating and Certificate Number of Technician or Repair Facility. (See back pages for other specific entries.)

YEAR 20____ DATE	RECORDING TACH TIME	TODAY'S FLIGHT	TOTAL TIME IN SERVICE	**Description of Inspections, Tests, Repairs, and Alterations** Entries must be endorsed with Name, Rating and Certificate Number of Technician or Repair Facility. (See back pages for other specific entries.)

YEAR 20 ____ DATE	RECORDING TACH TIME	TODAY'S FLIGHT	TOTAL TIME IN SERVICE	**Description of Inspections, Tests, Repairs, and Alterations** Entries must be endorsed with Name, Rating and Certificate Number of Technician or Repair Facility. (See back pages for other specific entries.)

YEAR 20_____ DATE	RECORDING TACH TIME	TODAY'S FLIGHT	TOTAL TIME IN SERVICE	**Description of Inspections, Tests, Repairs, and Alterations** Entries must be endorsed with Name, Rating and Certificate Number of Technician or Repair Facility. (See back pages for other specific entries.)

'EAR 0 _____ ATE	RECORDING TACH TIME	TODAY'S FLIGHT	TOTAL TIME IN SERVICE	**Description of Inspections, Tests, Repairs, and Alterations** Entries must be endorsed with Name, Rating and Certificate Number of Technician or Repair Facility. (See back pages for other specific entries.)

YEAR 20 ____ DATE	RECORDING TACH TIME	TODAY'S FLIGHT	TOTAL TIME IN SERVICE	**Description of Inspections, Tests, Repairs, and Alterations** Entries must be endorsed with Name, Rating and Certificate Number of Technician or Repair Facility. (See back pages for other specific entries.)

YEAR 20____ DATE	RECORDING TACH TIME	TODAY'S FLIGHT	TOTAL TIME IN SERVICE	**Description of Inspections, Tests, Repairs, and Alterations** Entries must be endorsed with Name, Rating and Certificate Number of Technician or Repair Facility. (See back pages for other specific entries.)

YEAR 20 _____ DATE	RECORDING TACH TIME	TODAY'S FLIGHT	TOTAL TIME IN SERVICE	**Description of Inspections, Tests, Repairs, and Alterations** Entries must be endorsed with Name, Rating and Certificate Number of Technician or Repair Facility. (See back pages for other specific entries.)

EAR O ____ ATE	RECORDING TACH TIME	TODAY'S FLIGHT	TOTAL TIME IN SERVICE	**Description of Inspections, Tests, Repairs, and Alterations** Entries must be endorsed with Name, Rating and Certificate Number of Technician or Repair Facility. (See back pages for other specific entries.)

YEAR 20 _____ DATE	RECORDING TACH TIME	TODAY'S FLIGHT	TOTAL TIME IN SERVICE	**Description of Inspections, Tests, Repairs, and Alterations** Entries must be endorsed with Name, Rating and Certificate Number of Technician or Repair Facility. (See back pages for other specific entries.)

EAR 0 ____ ATE	RECORDING TACH TIME	TODAY'S FLIGHT	TOTAL TIME IN SERVICE	**Description of Inspections, Tests, Repairs, and Alterations** Entries must be endorsed with Name, Rating and Certificate Number of Technician or Repair Facility. (See back pages for other specific entries.)

YEAR 20 ____ DATE	RECORDING TACH TIME	TODAY'S FLIGHT	TOTAL TIME IN SERVICE	**Description of Inspections, Tests, Repairs, and Alterations** Entries must be endorsed with Name, Rating and Certificate Number of Technician or Repair Facility. (See back pages for other specific entries.)

YEAR 20 ____ DATE	RECORDING TACH TIME	TODAY'S FLIGHT	TOTAL TIME IN SERVICE	**Description of Inspections, Tests, Repairs, and Alterations** Entries must be endorsed with Name, Rating and Certificate Number of Technician or Repair Facility. (See back pages for other specific entries.)

YEAR 20____ DATE	RECORDING TACH TIME	TODAY'S FLIGHT	TOTAL TIME IN SERVICE	**Description of Inspections, Tests, Repairs, and Alterations** Entries must be endorsed with Name, Rating and Certificate Number of Technician or Repair Facility. (See back pages for other specific entries.)

YEAR 20____ DATE	RECORDING TACH TIME	TODAY'S FLIGHT	TOTAL TIME IN SERVICE	**Description of Inspections, Tests, Repairs, and Alterations** Entries must be endorsed with Name, Rating and Certificate Number of Technician or Repair Facility. (See back pages for other specific entries.)

YEAR 20 _____ DATE	TOTAL TIME IN SERVICE	**Major Repairs and Major Alterations** FAA Form 337 Date or the Work Order Number and the Approving Agency

YEAR 20____ DATE	TOTAL TIME IN SERVICE	**Major Repairs and Major Alterations** FAA Form 337 Date or the Work Order Number and the Approving Agency

YEAR 20 ____ DATE	TOTAL TIME IN SERVICE	**Major Repairs and Major Alterations** FAA Form 337 Date or the Work Order Number and the Approving Agency

YEAR 20 ____ DATE	AD NUMBER	TOTAL TIME IN SERVICE	**Airworthiness Directives** Chronological Listing and Method of Compliance

YEAR 20 ____ DATE	AD NUMBER	TOTAL TIME IN SERVICE	**Airworthiness Directives** Chronological Listing and Method of Compliance

YEAR 20 _____ DATE	AD NUMBER	TOTAL TIME IN SERVICE	**Airworthiness Directives** Chronological Listing and Method of Compliance

YEAR 20 _____ DATE	TOTAL TIME IN SERVICE	**Manufacturers' Mandatory Service Bulletins** Chronological Listing and Method of Compliance

YEAR 20 ____ DATE	TOTAL TIME IN SERVICE	**Equipment Addition, Removal or Exchange** Item / Manufacturer's Name / Model / Serial Number	
			☐ Addition of Optional Equipment ☐ Removal of Optional Equipment ☐ Addition of Required—Exchanged for Optional ☐ Removal of Required—Exchanged for Optional
			☐ Addition of Optional Equipment ☐ Removal of Optional Equipment ☐ Addition of Required—Exchanged for Optional ☐ Removal of Required—Exchanged for Optional
			☐ Addition of Optional Equipment ☐ Removal of Optional Equipment ☐ Addition of Required—Exchanged for Optional ☐ Removal of Required—Exchanged for Optional
			☐ Addition of Optional Equipment ☐ Removal of Optional Equipment ☐ Addition of Required—Exchanged for Optional ☐ Removal of Required—Exchanged for Optional
			☐ Addition of Optional Equipment ☐ Removal of Optional Equipment ☐ Addition of Required—Exchanged for Optional ☐ Removal of Required—Exchanged for Optional
			☐ Addition of Optional Equipment ☐ Removal of Optional Equipment ☐ Addition of Required—Exchanged for Optional ☐ Removal of Required—Exchanged for Optional
			☐ Addition of Optional Equipment ☐ Removal of Optional Equipment ☐ Addition of Required—Exchanged for Optional ☐ Removal of Required—Exchanged for Optional
			☐ Addition of Optional Equipment ☐ Removal of Optional Equipment ☐ Addition of Required—Exchanged for Optional ☐ Removal of Required—Exchanged for Optional
			☐ Addition of Optional Equipment ☐ Removal of Optional Equipment ☐ Addition of Required—Exchanged for Optional ☐ Removal of Required—Exchanged for Optional

YEAR 20 ____ DATE	TOTAL TIME IN SERVICE	**Equipment Addition, Removal or Exchange** Item Manufacturer's Name Model Serial Number	
			☐ Addition of Optional Equipment ☐ Removal of Optional Equipment ☐ Addition of Required—Exchanged for Optional ☐ Removal of Required—Exchanged for Optiona
			☐ Addition of Optional Equipment ☐ Removal of Optional Equipment ☐ Addition of Required—Exchanged for Optional ☐ Removal of Required—Exchanged for Optiona
			☐ Addition of Optional Equipment ☐ Removal of Optional Equipment ☐ Addition of Required—Exchanged for Optional ☐ Removal of Required—Exchanged for Optiona
			☐ Addition of Optional Equipment ☐ Removal of Optional Equipment ☐ Addition of Required—Exchanged for Optional ☐ Removal of Required—Exchanged for Optiona
			☐ Addition of Optional Equipment ☐ Removal of Optional Equipment ☐ Addition of Required—Exchanged for Optional ☐ Removal of Required—Exchanged for Optiona
			☐ Addition of Optional Equipment ☐ Removal of Optional Equipment ☐ Addition of Required—Exchanged for Optional ☐ Removal of Required—Exchanged for Optiona
			☐ Addition of Optional Equipment ☐ Removal of Optional Equipment ☐ Addition of Required—Exchanged for Optional ☐ Removal of Required—Exchanged for Optiona
			☐ Addition of Optional Equipment ☐ Removal of Optional Equipment ☐ Addition of Required—Exchanged for Optional ☐ Removal of Required—Exchanged for Optiona
			☐ Addition of Optional Equipment ☐ Removal of Optional Equipment ☐ Addition of Required—Exchanged for Optional ☐ Removal of Required—Exchanged for Optiona

Notes

Notes

Notes

Notes